6TH GRADE
MATHEMATICS

Africa
World Books
Pty Ltd

The publisher wishes to acknowledge and thank Dr. Douglas H. Johnson for his invaluable help and support for Africa World Books and its mission of preserving and promoting African cultural and literary traditions and history. Dr. Johnson and fellow historians have been instrumental in ensuring that African people remain connected to their past and their identity. Africa World Books is proud to carry on this mission.

Cover design, typesetting and layout: Africa World Books
Unit 3, 57 Frobisher St, Osborne Park, WA 6017
P.O. Box 1106 Osborne Park, WA 6916

Unit 1. Understanding and Applying Whole Numbers

Q1.

Write 5, 921,004 in word form

   A. Five million nine hundred twenty-one thousand four

   B. Five hundred ninety -two thousand one hundred four

   C. Five million nine hundred twenty-one

   D. Five hundred thousand nine hundred two hundred four

Q2. Use <, > or = to compare the following whole numbers.

8,945 _____ 8,895

A.   <

B.   >

C.   =

D.   More information is needed.

Q3. A boda boda operator in Juba makes 7 trips in a day. How much does he make in the   month of July if he charges 150 SSPs a trip?

A. 1,050 SSPs.

B. 210 SSPs.

C. 31,500 SSPs.

D. 32, 550 SSPs.

Use the bar graph below to answer question 4

Facebook Friends

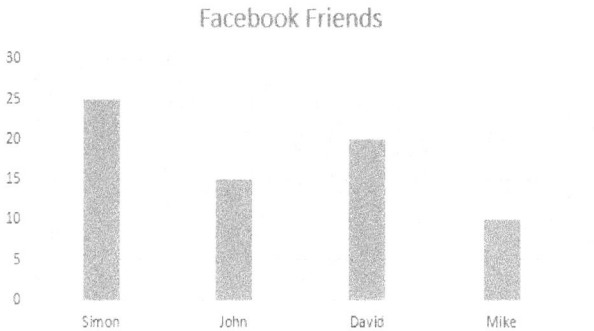

Q4.

How many more Facebook friends does Simon have than Mike?

A. 15

B. 25

C. 10

D. 250

Q5.

South Sudan Supreme plane carries 150 passengers per a trip. In a week, it carries a total of 900 passengers. How many trips does the plane make in a week?

A. 1, 050 trips

B. 750 trips

C. 6 trips

D. 60 trips

Q6.

Arrange the following whole numbers in ascending order

569, 705, 472, 891

A.   891, 472, 705, 569

B.   472, 569, 891, 705

C.   472, 569, 705, 891

D.   472, 705, 569, 891

Q7.

There are 56, 401 pupils in Royal Academy.
There are 45, 091 pupils in Alliance Academy.
How many students are in 2 academies?

A.   11, 310

B.   101, 492

C.   101, 500

D.   11,300

Q8.

Estimate the product of 685 and 78.
   A.   42,000

   B.   68,500

   C. 56,000

   D. 607

Q9.
   Round 67, 995 to the nearest tens
   A.   67,000
   B.   67, 990
   C.   68,900

   D. 68,00

Q10.

Manyang caught a total of 705 kgs of tilapia. If an average weight of each tilapia is 5 kgs, how many pieces of tilapia did he catch?

A. 141 pieces of tilapia

B. 700 pieces of tilapia

C. 800 pieces of tilapia

D. 3,525 pieces of tilapia

Q11.

Find the product of 452 and 231.

A. 2

B. 221

C. 104, 412

D. 683

Q12.

$35^2 =$

A. 1,225

B. 33

C. 70

D. 37

Q13.

$\sqrt{625} =$

A. 313

B. 312

C. 25

D. 1,250

Q14.

Which of the following numbers is Not a perfect square?

A. 16

B. 9

C. 900

D. 84

Q15.

Which of the following numbers is divisible by both 8 and 11?

A. 121

B. 22

C. 88

D. 19

Q16.

The White Nile river is about 6,650 Kilometers long. The Amazon river is about 6,400 Kilometer long. How much longer is White Nile river than Amazon river?

A. 12, 900 km longer

B. 250 km longer

C. 12, 000 km longer

D. 200 km longer

Q17.

Find the sixth number in the pattern below
1, 4, 9, …

A. 16

B. 25

C. 49

D. 36

Q18.

Estimate the quotient of 589 and 12

    A. 60

    B. 610

    C. 50

    D. 577

Q19.

The population of South Sudan according 2010 census is 8,736,939. The 3 in the thousands place is____times the value of 3 in the tens place.

    A. 100

    B. 3,000

    C. 1,000

    D. 300

Q20.

Use the <, > or = to compare the following sides.

Side A   Side B

$\sqrt{81}$ the product of 4 and 3

A.   Side A > Side B

B.   Side A < Side B

C.   Side A = Side B

D.   More information is needed.

Unit 2. Fractions, Decimals and Percentages

Q1. $\frac{3}{5} + \frac{1}{3} =$

A. $\frac{4}{8}$

B. $\frac{2}{8}$

C. $\frac{4}{15}$

D. $\frac{14}{15}$

Q2. $\frac{1}{2} - \frac{1}{3} =$

A. 0

B. $\frac{2}{5}$

C. $\frac{1}{6}$

D. $\frac{5}{6}$

Q3.

$5 \div \dfrac{1}{2} =$

  A. $2\dfrac{1}{2}$

  B.   10

  C. $\dfrac{1}{10}$

  D. $\dfrac{2}{5}$

Q4.

Each student in 6th grade D in Royal Preparatory gets $\frac{1}{8}$ of a pizza. There are 3 full pizzas . How many students are in 6th grade D?

  A. $3\dfrac{1}{8}$

  B.   24

  C.   11

  D.   5

Q5. Akau makes $120\frac{3}{4}$ SSPs. per hour as a waiter at Mama Achol's Restaurant in Bortown. He works for 8 hours in a day. How much does Akau earn in a day?

A. 966 SSP.

B. 72 SSP.

C. 255 SPP.

D. 120.75 SSP.

Q6. Akau washes dishes in 8 minutes. Kuchkon washes the same number of dishes in 5 minutes. If Akau and Kuchkon wash dishes together, how long will it take them ?

   A.   3 minutes

   B.   40 minutes

   C.   13 minutes

   D. $3\frac{1}{13}$ minutes

Q7. Achol cycles $4\frac{1}{2}$ km from home to school. If it takes her $2\frac{1}{2}$ minutes, what is her average speed? (hint: average speed $= \frac{distance}{time}$)

A. $1\frac{4}{5}$ km/min.

B. $\frac{5}{9}$ km/min.

C.   7km/min

D.   2 km/min

Q8.

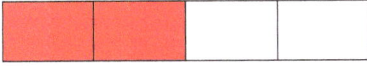

Red Ribbon

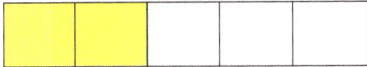

Yellow Ribbon

How much longer is red ribbon than yellow ribbon?

A.   0

B. $\dfrac{1}{10}$

C. $\dfrac{1}{6}$

D. $\dfrac{2}{5}$

Q9.

$$\frac{1}{20} \div 1\frac{1}{4} =$$

A. $\frac{4}{15}$

B. $\frac{1}{5}$

C. $\frac{1}{25}$

D. 1

Q10.

What is the reciprocal of 3/5?

A. $\frac{2}{5}$

B. $1\frac{2}{3}$

C. 2

D. 15

Q11.

Write 1.35 in word form?

A. One hundred thirty -five

B. One and thirty-five

C. One and thirty-five hundred

D. One and thirty-five hundredths

Q12.

Convert 5/8 to decimal. Round your answer to thousandths place if necessary.

A. 0.625

B. 1.6

C. 0.375

D. 2.667

Q13.

Convert 1/3 to decimal

A. $0.\overline{3}$

B. $0.\overline{6}$

C. 3

D. $0.\overline{7}$

Q14.

$12.356 \times 0.5 =$

A. 617.8

B. 0.6178

C. 61.78

D. 6.178

Q15.

$18.625 \div 0.25 =$

A. 7.45

B. 74.5

C. 0.745

D. 745

Q16.

Round 409.1589 to the nearest thousandths

    A.   400

    B.   409.16

    C.   409.159

    D.   410

Q17.

Lado ran around the football field in 7.45 minutes. Taban ran around the football field in 5.69 minutes. How much longer did Lado take than Taban?

A. 13.14 minutes

B. 6.435 minutes

C. 7.3931 minutes

D. 1.76 minutes

Q18.

Akau walked 12.75 km in 5 days. How many kilometers did he walk per day?

A. 17.75 km

B. 7.75 km

C. 2.55 km

D. 25.5 km

Q19.
Nakuma spent $0.75 on legemat. Nyapethe spent 1/5 of a dollar on legemat. Who spent more on legemat? Explain your reasoning.

A. Nakuma, because $0.75 is the same as ¾ of a dollar which is greater than 1/5 of a dollar.

B. Nyapethe, because 1/5 of a dollar is the same as $5.00 which is greater than $0.75.

C. They both spent about the same amount because $0.75 is the same as 1/5 of a dollar.

D. More information is needed to compare $0.75 and 1/5 of a dollar.

Q20.

Find the quotient and express your answer as a fraction.

$0.625 \div 125 =$

    A. 1/200

    B. 1/20

    C. ½

    D. 1/2000

Q21.

Convert $0.\bar{1}$ to fraction.

    A. 11/100

    B. 11/10

    C. 1/9

    D. 9/11

Q22.

Kuol got 7 out of 9 Math problems correct. Express his grade as percent. Round your answer to a tenth of percent.

A. 77.8 %

B. 77.778 %

C. 128.6 %

D. 12.86 %

Q23.

A birthday cake is sold at 25 % off the regular price. If Mama Akau paid $25 after the discount, what is the regular price? Round your answer to the nearest dollar.

A. $31.25

B. $50.00

C. $6.25

D. $33

Q24.

There are 457 students at Gudele Elementary school. If 235 of the students are girls, what percent of the students is boys? Round your answer to a hundredth.

A. 51.42 %

B. 51.4%

C. 48.58 %

D. 48.6%

Q25.

At Malakal school system Spelling Bee, Ujuko spelled 12 out of 17 words correctly, while Jalduong correctly spelled 72 % of the words. Who won the Spelling Bee? Explain your reasoning.

A. Ujuko won the Spelling Bee because 12/17 is about 71 % which is more than 72%.

B. Jaldong won the Spelling Bee because 12/17 is about 71% which is less than 72%.

C. Both won because 12/17 is about 72% which is equal to 72%

D. More information is needed to determine who won.

Q26.

23 % of 2,300 SSPs. =

A. 529 SSPs.

B. 5290 SSPs.

C. 52.90 SSPs

D. $1.85

Q27.

Express 56 % as a decimal

A.   5.6

B.   0.56

C.   0.056

D.   56.0

Q28.

Girls in Royal Academy represent 56 % of student population. There are 440 boys. How many girls are in Royal Academy?

A.   560

B.   57

C.   440

D.   1,000

Q29.

72 % of the revenue collected last year in Marol Market was used for teachers' salaries. The rest of the revenue was used for other projects in Bortown Municipality. Find the total revenue collected, if 4,200,000 South Sudanese Pounds was used for other project?

A.   19,200,000 SSPs.

B.   15,000,000 SSPs.

C.   10,800,000 SSPs.

D.   25,800,000 SSPs.

Q30.

Limuro gave his customers 5.6 % discount on a kilogram of simsim. What is the original price of 1 kg of simsim if the discount is 28 SSPs.?

A. 560 SSPs.

B. 28 SSPs.

C. 500 SSPs.

D. 528 SSPs.

NOTES

Unit 3. Ratio and Proportion

Q1.

There are 35 girls and 14 boys in 6th Grade A at Kiir Academy. Express the ratio of boys to girls in simplest form.

A.   5:2

B.   2:5

C.   14:35

D.   35:14

Q2. At Juba University Teaching Hospital, the ratio of female doctors to male doctors is 1:3. If there are 45 male doctors, what is the total number of doctors at Juba University Teaching Hospital?

A. 135

B. 15

C. 60

D. 180

Q3. At Yambio Medical College, the number of female students is directly proportional to the number of male students. In class of 2011, there were 25 female students and 35 male students. If there were 70 male students in class of 2019, how many students were enrolled in 2019?

A. 60

B. 50

C. 70

D. 120

Q4.

$3:5 = 21:?$

$? =$

A. 26

B. 24

C. 35

D. 56

Q5.

Obour's family decided to save 2SSPs for every 5SSPs they spend. In one year, they realized that their expenditure was 700,000 SSPs.

Part A. How much was their total income that year?

Part B. How much did they save that year?

Part C. If 35 % and 8 % of their savings is respectively spent on Master Obour's schooling and Ms. Obour's wardrobes, how much money is in their savings account? (Assume that no interest is earned on their savings)

Q6.

To prepare Asida, you need 3 cups of sorghum flour for every 4 cups of water. Regina, in preparation for her daughter's wedding, she used 45 cups of Sorghum flour for Asida. How many cups of water did she use?

A. 15

B. 12

C. 60

D. 52

Use the pie chart below to answer question 7

FAVORITE BEVERAGE

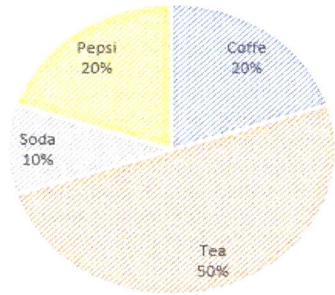

Q7.
If 24 people have soda as their favorite beverage, how many people like coffee and Pepsi?

A. 48

B. 96

C. 40

D. 240

Q8. In Raja Primary school the ratio of boys to girls is 7:3. There are 800 students in the school. During the Spelling Bee competition $1/10$ of girls and $1/8$ of boys took part. How many students took part in the Spelling Bee competition?

A. 94

B.   24

C. 240

D. 560

Q9.

Simplify 0.45: 0.09

   A.   1:5

   B.   5:1

   C.   9:45

   D.   45:9

Q10.

Linda agreed to pay her son, Lado 5 SSPs for every 2 Math problems he correctly answered on his homework. Lado received a total of 50 SSPs.

Part A. How many Math problems did Lado answer correctly?

Part B. If there were 25 problems assigned for the homework, what percent of the homework did he correctly answer?

Part C. Lado promised his mother to cut the lawn for every 8 problems he missed. Did he miss enough problems to cut the lawn? Explain your reasoning.

Q11.

$$\frac{y}{65\,\%} = \frac{1.40}{2.60}$$

A. 35%

B. 26 %

C. 14%

D. 61%

Unit 4. Measurement: Length, Area, Volume, Capacity, Weight, Money, Rates, Time and Speed

Q1.

Find the circumference of the circle below. $\pi = \dfrac{22}{7}$

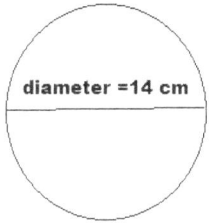

diameter =14 cm

A. 154 cm^2

B. 44 cm^2

C. 154 cm^2

D. 44 cm^2

Q2.

Mama Anya Anya has a circular shaped kitchen garden. The kitchen garden's diameter is 56 meters. She plans to put a fence around its circumference to keep away rodents.

Part A.

How many meters of fencing will she need? ($\pi = \frac{22}{7}$)

Part B.

If a meter of fencing costs 50 SSPs, how much will she spend on fencing the entire kitchen garden?

Part C.

If the cost of fencing the garden represents 4.4 % of her salary, how much is her salary?

Q3.
Find the area of the shaded region.

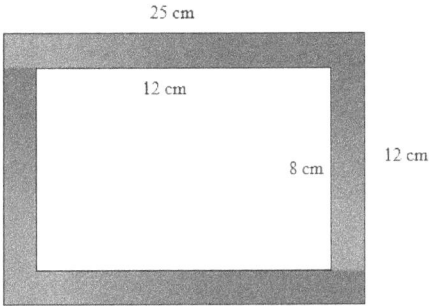

25 cm

12 cm

12 cm

8 cm

A. 396 cm^2

B. 300 cm^2

C. 204 cm^2

D. 96 cm^2

Q4.

Use the figure below to answer parts A and B.

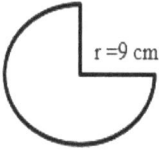

r = 9 cm

Part A. Find the perimeter of the figure. Express your answer in terms of π.

A. $81\pi \text{ cm}^2$

B. $9\pi \text{ cm}$

C. $\dfrac{22}{7}\pi \text{ cm}$

D. $\left(\dfrac{22}{7}\pi + 18\right) cm$

Part B. Simplify and round your answer from part A to the nearest thousandths. $\pi = \dfrac{22}{7}$

Q5.

Find the area of the figure ABCDE below, given AB = CD = 8 m, AC = BD = 6 m, and CE = 12 m.

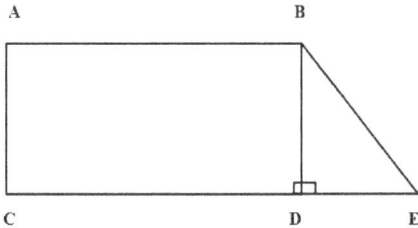

A. 48 m^2

B. 12 m^2

C. 60 m^2

D. 36 m^2

Q6.

Us the figure below to answer parts A, B and C.

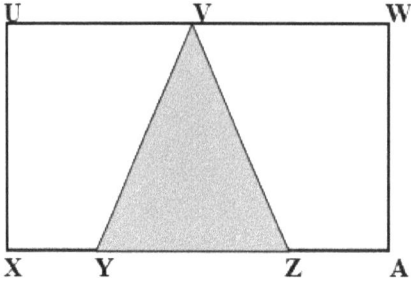

Part A. Given UW = XA = 15 m, UX = AW = 9 m and YZ = 9 m, find the area of the unshaded region.

Part B. What percent of UWAX is shaded?

Part C. Convert your answer from Part A to hectare **(hint: 1 ha = 10,000 m²).**

Q7.

Use the solid below to answer parts A, B and C.

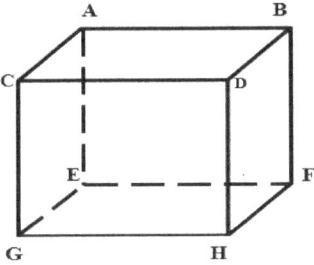

Part A. Find the volume of cube above, given
AB=AC=BF = 9 m

Part B. The Department of Agriculture and Forestry at John Garang Memorial University of Science and Technology used the cube above as an aquarium. If each fish requires a space of 50 cm³, how many fish will be accommodated in the aquarium?

Part C. Given that the mass of the cube above is 53kg, calculate its density. If necessary, round your answer to the nearest hundredths.

Part D. If Manyang, the student in charge of the aquarium uses a 45- cubic -meter jerrican to fetch water from the White Nile River to fill the aquarium, how many trips will he make?

Q8.

Nyandeng, a local baker in Bortown bakes cuboid-shaped bread in dimensions of 3 cm by 5 cm by 6 cm.

Part A. She transports her loaves of bread in a carton box with dimensions of 12 cm by 8 cm by 9 cm. How many loaves of bread can she fit in the box?

Part B. She makes and supplies 12 boxes per day. How much does she make in the month of July if each box costs 300 SSPs?

Q9.

Jl conference table is shaped as below.

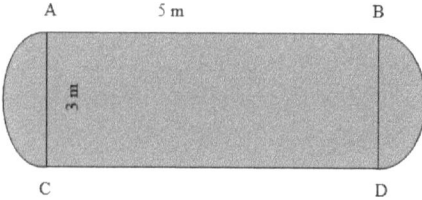

Cabinet members are required be seated 1. 5 m apart due to Coronavirus social distancing. How many cabinets will be seated around the conference table at a time? $\pi \cong 3.14$

Q10.

Given the area of unshaded region is 48 cm^2, find the length of wz.

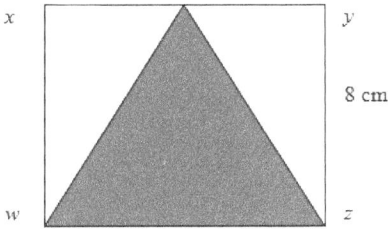

A. 6 cm

B. 12 cm

C. 40 cm

D. 36 cm

Q11.

Yau Yau is a commercial beekeeper on Buma Plateau. In 2019, he harvested 240 liters of honey.

Part A. He put his honey into a 500 ml bottle. How many full bottles of honey did he make that year?

Part B. He sold each bottle to his local customers at 300 SSPs. How much did he make from the sale of his honey that year?

Part C. If the Buma municipality levied a sale tax of 50 SSPs for every sale of 1,000 SSPs, what was Yau Yau's net sale that year?

Part D. If Yau Yau's labor and cost of the materials amounted to 98,000 SSPs, did he make profit or loss? Find either profit or loss he made that year.

Q12.

South Sudan government donated 10 tons of sorghum flour to the victim of flooding in Jonglei State.

Part A. If a family size of 5 received 50 kg-bag of sorghum flour, how many families were affected by the flooding?

Part B. If the government, after getting 30 % discount paid a total 3,500,000 SSPs, what was the original price per bag?

Part C. The price of a 50kg-bag of sorghum flour at Marol Market cost 300 SSPs. Did the government make the right decision to buy and transport the donation from Reng? Explain your reasoning.

Q13.

Agut Aja bought a bull for 120,000 SSPs and sold it for $150,000 SSPs. Find his percent profit or loss.

A. 25 % profit

B. 25 % loss

C. 75% profit

D. 75% loss

Q14.

Dau bought 5 boxes of mango fruit @ 350 SSPs. 1 box of the fruit was rotten, and he sold the remaining boxes @245 SSPs. What is his percentage loss?

A. 44 %

B. 77 %

C. 17.50 %

D. 98 %

Q15.

Alier bought a basket of thou at 45% discount. If he paid 900 SSPS after discount, how much was the original price of the basket of thou?

A. 1,100 SSPs.

B. 900 SSPs.

C. 2,000 SSPs.

D. 4,500 SSPs.

Q16.

The graph below shows a journey by matatu from Juba to Rumbek. Use it to answer the following question. What is the matatu's average speed between Juba and Rumbek?

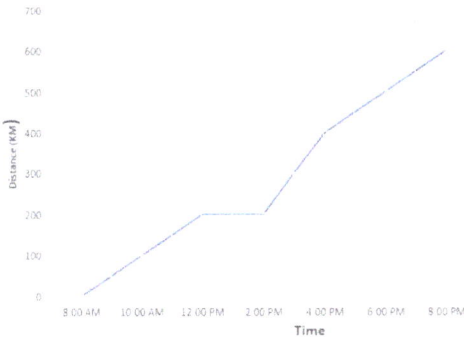

A. 60 km/hr.

B. 200 km/hr.

C. 50 km/hr.

D. 600 km/hr.

Q17.

Aleer and his family drove from Duk Padiet to Juba for a total of 15 hours. If he drove at average speed of 75 km/hr., how far apart are the two places?

A. 11,250 km.

B. 90 km.

C. 75 km

D. 250 km.

Q18.

If Aleer and his family left Duk Padiet on Monday at 3.00 PM and stopped in Mar and Bortown to gas up and dined for a total of 2 hrs., at what time did they arrive in Juba?

A. 8.00 PM on Monday

B. 8.00 AM on Monday

C. 8.00 PM on Tuesday

D. 8.00 AM on Tuesday

Q19.

Achala cycles at $\frac{4}{3}$ m/s. to a school which is 800 meters away from her home. How long does it take her to get to school? Write your answer in minutes

A. 10 min.

B. 60 min.

C. 600 min.

D. 1,066 min.

Q20.

If Achala's first period starts at 7.30 AM., what time should she leave her home to get school in time for her lesson?

A. 7. 30 AM.

B. 7.40 AM.

C. 7.20 AM

D. 7.50 AM.

Unit 5. Geometry

Q1.

Find the measure of x:

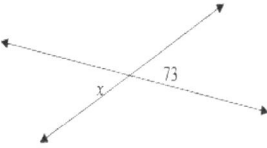

A. 107^0

B. 73^0

C. 180^0

D. 287^0

Q2.

Use a pair of compasses and a rule to construct angle of 90^0.

Q3.

Use a pair of compasses and a rule to construct angle of 60^0.

Q4.
Use a pair of compasses and a rule to construct angle of 120^0.

Q5.
Use a pair of compasses and a rule to construct angle of 30^0

Q6.

Use a pair of compasses and a rule to construct angle of 75^0

Q7.

Use a pair of compasses and a rule to construct angle of 15^0

Q8.

Vertically opposite angles are congruent.

A. True

B. False

C. They do not have any relationship.

D. They are similar.

Q9.

Use a pair of compasses and rule to construct a circle whose radius is 3 cm. Label the radius and diameter of the circle.

Q10.

If angles P and R are vertically opposite angles, which of the following is always true?

A. The measures of angles P and R are equal.

B. Both angles are complementary.

C. Measure of angle P is greater than measure of angle R

D. Vertically opposite angles are supplementary.

NOTES

Unit 6. Algebra

Q1.

Abuk bought 3 dried fish and 2gs of kombo for a total of 800 SSPs.

Part A. If 2gs of kombo cost 60 SSPs., write an equation for the total cost of 3 dried fish and 2gs of kombo. Let f be the price of a dried fish.

Part B. Solve the equation above to find the cost of 1 dried fish.

Q2

The length of a rectangle is twice its width.

Part A. Write the equation representing perimeter.

Part B. Find the values of length and width if the perimeter is 48 cm.

.

Q3.

The sum of a number and five is greater than 10.

Part A. Write the inequality representing the situation above.

Part B. Solve the inequality above for the value of unknown number

Q4.

$$\frac{1}{2}x + 4 \geq 2$$

A. $(-\infty, -4]$

B. $[-4, \infty)$

C. $(-\infty, -4]$

D. $(-4, \infty)$

Q5.

Atem bought 3kgs of thou @500 SSPs and a bag of sorghum flour. If the total cost of 3kgs.of thou and a bag of sorghum flour is at most 30,000 SSPs., find the cost of bag of sorghum flour? Let b be the price of a bag of sorghum flour.

Q6.
Given $x + 12 = 80$, solve for x.

A. 68

B. 92

C. 13.3

D. 960

Q7.

Simplify $6x + 4y - 3x + y$

A. $6x$

B. $3x+5y$

C. $7x + y$

D. $10x - 2y$

Q8.

Simplify $\frac{x}{8} + \frac{x}{3}$

A. $\frac{2x}{11}$

B. $\frac{11x}{24}$

C. $2\frac{2}{12}x$

D. $\frac{x}{5}$

Q9.
Solve for k:

$$\frac{k}{4} + 5 = 12$$

A. 68

B. 28

C. 8

D. 9

Q10.

Given $x = 2$ and $y = 3$, evaluate the expression below.

$$(x + y)^2$$

A. 10

B. 1

C. 25

D. 13

Q11.

Given -3y ≤ 15, what is the solution set of y?

A. $(-\infty, -5)$

B. $(-\infty, -5]$

C. $[-5, \infty)$

E. $[-\infty, -5]$

Q12.

A number plus its 35 % is equal to 135. What is the number?

A. 35

B. 65

C. 100

D. 200

Q13.

Given x = 3 and y = 2, find value of the expression below.

$$y^x$$

A. 6

B. 9

C. 5

D. 8

Q14.

Which of the following numbers is not in the solution set of $k \leq 3$?

A. 2

B. 5

C. -1

D. 0

Q15.

Find the solution set of:

$10 \leq x + 2 \leq 18$.

A. $[8, 16]$

B. $[12, 20]$

C. $(8, 16]$

D. $[8, 16)$

Q16.

The sum of 3 consecutive odd numbers is 21. Find the 3 odd numbers.

Q17.

The perimeter of the figure PQRS is 62.8 cm. Find the value x. Round your answer to the nearest ten thousandths. $\pi \cong 3.14$

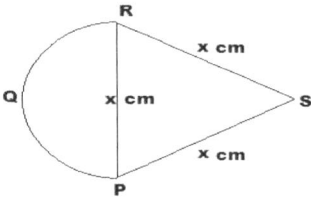

Q18.

$\angle R$ and $\angle S$ are complementary angles. If m$\angle R$ is 18^0 more than m$\angle S$, find $m\angle S$ and $m\angle R$.

Q19.

Alier & Sons canoeing company on White Nile River charges 500 SSPs down payment plus 50 SSps per an hour for a canoe rental. Angelo Mou wanted to spend at most 2,000 SSPs, for how long did he take the canoe?

Q20.

MTN cellphone company charges 30 SSPs per a minute and 600 SSPs down payment. Zain cellphone company charges 10 SSPs per a minute and 800 SSPs down payment.

Part A. Which company offers a cheaper service?

Part B. How many minutes of calls will the two plans be equal?

Q21.

Given $x = 5$ and $p = 2$, use $<$, $>$, or $=$ compare the following expressions.

$$x^y \underline{\hspace{1cm}} y^x$$

A. $<$

B. $>$

C. $=$

D. More information needed.

Q22.

The number of girls in 6th grade at Bor B is 8 more than the number of boys. If there are 26 pupils in 6th grade, how many boys are there?

A. 17

B. 9

C. 8

D. 34

Q23.

Area of Δikl below is 128 cm². Find value of x

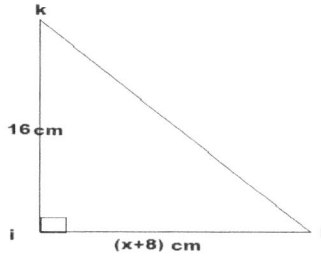

A. 24 cm

B. 8 cm

C. 16 cm

D. 128 cm

Q24.

Given $AB = BC = CE = x$ and the volume of the cube is 2^3 cm^3, find the value of x.

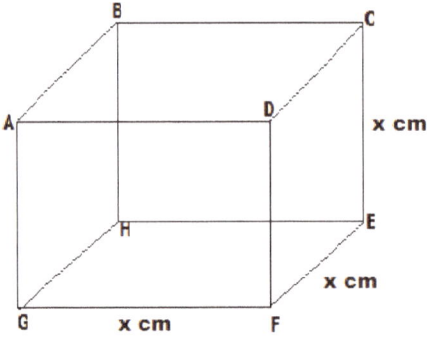

Q25.

Which of the following shows the solution set of -$8 < 5x + 2 \leq 27$?

A.

B.

C.

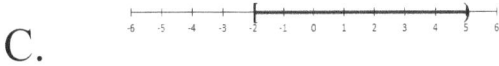

D.

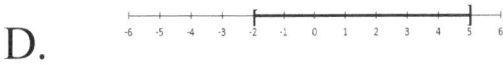

Q26.

Find the solution set of $x + 5 > 2$.

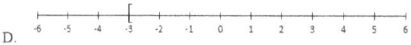

A.

B.

C.

D.

Q27.

Solve for x: $3x + 5 = 3x + 19$.

A. 0

B. 8

C. Infinitely many solutions

D. No solution

Q28.

Solve for x: $18x + 5 = 18x + 5$

A. 0

B. $\dfrac{5}{12}$

C. Infinitely many solutions

D. No solution

Q29.

Show the solution set of
$-16< (4x + 4) \leq 24$ on the number line below

```
 ├──┼──┼──┼──┼──┼──┼──┼──┼──┼──┼──┼──┤
 -6  -5  -4  -3  -2  -1   0   1   2   3   4   5   6
```

NOTES

Unit 7. Statistics

Use the circle graph below shows types of vegetables that Nyibol harvested from her garden in 2020. Use the circle graph to answer questions 1-3

NUMBER OF VEGETABLES

Okra   Potato   Kudhura   Egg Plants

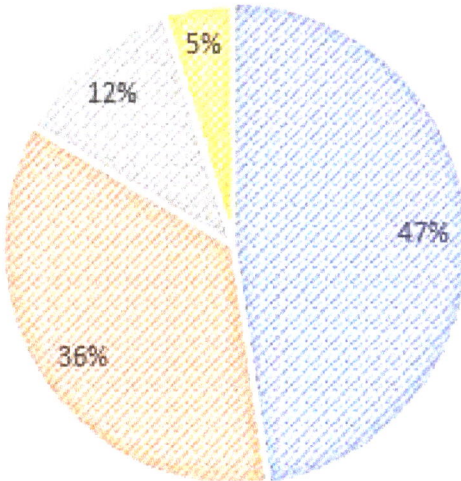

5%

12%

47%

36%

Q1.

If Nyibol harvested 90 kgs of potatoes in 2020, what is the total kgs of vegetables did she harvest in 2020?

A. 25 kgs

B. 250 kgs

C. 360 kgs

D. 180 kgs

Q2.

How more kgs of okra did Nyibol harvest than kudhura?

A. 87.5 kgs

B. 12 kgs

C. 147.5 kgs

D. 59 kgs

Q3.

If 1kg of potatoes, 1 kg of eggplants, 1kg of kudhura and 1 kg of okras respectively cost 125 SSPs., 75 SSPs., 200 SSPs., and 350 SSPs. How much did Nyibol make that year?

Pupils in 6th grade in Anyang Aluong Academy were surveyed about their favorite sports and the data collected is in the table below. Use the data in the table below to answer questions 4 – 5.

Type of sports	Number of students
Basketball	8
Football	9
Table Tennis	1
Swimming	3
Volleyball	4

Q4. Find the percentage of each sport.
Basketball:

Football:

Table Tennis:

Cross Country:

Volleyball:

Q5.

Draw a circle graph (pie chart) using the percentages from question #4.

The table below shows food types and their prices (SSPs) at Mama Keigi's restaurant in Yei. Use the table to answer question 6- 7.

Sour Milk	Awalwala	Asida	Rice	Kisera	Akow
	250 SSPs	150 SSPs	100 SSPs	200 SSPs	120 SSPs
Cassava Leaves stew	100 SSPs	300 SSPs	200 SSPs	350 SSPs	300 SSPs
Dama	150 SSPs	250 SSPs	250 SSPs	400 SSPs	250 SSPs
Chicken	150 SSPs	350 SSPs	300 SSPs	300 SSPs	350 SSPs
Kamunia	150 SSPs	200 SSPs	250 SSPs	300 SSPs	400 SSPs
Bamia	150 SSPs	150 SSPs	150 SSPs	300 SSPs	200 SSPs

Q6.

Mondi with his wife Nakuma and 2 sons went to Mama Keigi's Restaurant. How much did they spend if the following are their orders?

Mondi: Chicken with Rice

Nakuma: Cassava leaves with Asida

Son1: Sour milk with Awalwala

Son 2: Bamia with Kisera

Q7.

Modi's family gave 4 % tips after 2 % local tax . How much in total did Modi's family pay?

Q8.

Ajah scored an average of 83 % on her final exams in 6th grade. Find her percentage score in Mathematics if the following shows her performance in other subjects:

Mathematics: ?

Science: 82 %

Agriculture: 80 %

Home Science: 78 %

CRE: 90 %

Business Education: 78 %

Arabic: 80 %

Social Studies: 84 %

    A. 78 %

    B. 92 %

    C. 100 %

    D. 46 %

Use the circle graph below to answer question # 9.

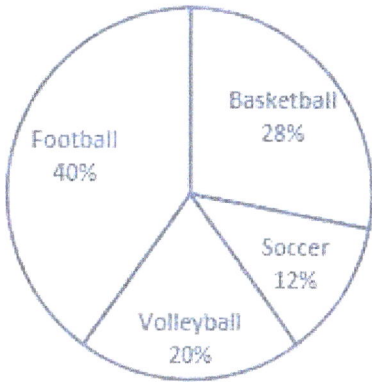

If 8 students played volleyball, how many students in total participated in the sporting activities?

A. 100

B. 32

C. 40

D. 80

Use the bar graph below to answer question #10

Facebook Friends

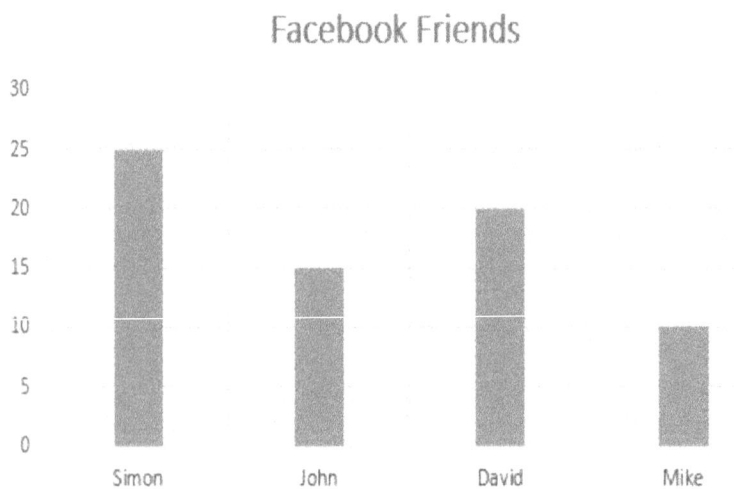

Q10.

How many less Facebook friends does Alier have than Ojukwu?

    A. 500

    B. 4,500

    C. 4,000

    D. 5,000

The table below shows the rainfall collected from March to August. Use it to answer questions 11-12.

Month	Rainfall (10ml)
March	45
April	50
May	35
June	20
July	10
August	5

Q11.

Find the median of the data above.

A. 275 ml

B. 500 ml

C. 350 ml

D. 250 ml

Q12.

Find the mean of the rainfall collected from March to August.

A. 275ml

B. 200ml

C. 350ml

D. 500ml

NOTES

www.ingramcontent.com/pod-product-compliance
Lightning Source LLC
Chambersburg PA
CBHW071715210326
41597CB00017B/2498